NEXTGEN LEARNERS PRESENTS

Think Like a Brain!™ Collection

Perry the Parietal & The Body Map Bash

Knowing Where Your Body Is and How It Moves

BOOK 6

Think Like a Brain!™
Collection Overview

Think Like a Brain!™ is a wildly engaging, silly-smart children's book collection that invites kids to explore how their brains work from the inside out. Through playful characters, imaginative settings, and rhythm-rich storytelling, this collection teaches neuroscience, social-emotional learning (SEL), and self-understanding in ways children absorb without even realizing they are learning.

Instead of telling children how to behave or what to feel, Think Like a Brain!™ shows them why their brains respond the way they do. Each story transforms complex brain science into friendly characters and relatable moments, helping kids feel capable, curious, and empowered rather than confused or ashamed.

Designed for children first and embraced by parents and educators as preventative education, this collection supports:

- Emotional regulation and self-awareness
- Executive functioning skills
- Neurodivergent understanding and inclusion
- Brain-body connection and growth
- Compassion for self and others

With humor, heart, and factual accuracy at its core, Think Like a Brain!™ helps children build the skills many adults wish they had learned sooner, one playful page at a time.

One of the collection's most powerful features is how the characters become real-life reminders for children long after the story ends. Once kids learn who the characters are and what jobs they do, they can begin to recognize those same processes in themselves. A big feeling might become, "It looks like Amy is trying to tell you something," or a moment of pause might sound like, "What does your Fred think?" By naming experiences through familiar characters, children learn to notice what is happening inside their brains, reflect on it, and move forward more quickly with understanding rather than frustration.

Series 1:
The Brainy Bunch (Brain Anatomy)

Series Overview: Meet the brain's core parts as a friendly team. Kids learn what each part does and how they work together to help thinking, feeling, moving, and learning.

1. **The Brainy Bunch**
   Focus: Brain overview and teamwork
2. **Brainstem Benny & The Automatic Adventure**
   Focus: Breathing, heartbeat, survival basics
3. **Bella Cerebellum's Balance Ballet**
   Focus: Balance, coordination, practice
4. **Olly Occipital & The Lookout Tower**
   Focus: Vision, visual processing
5. **Tammy Temporal & The Soundy Sound Storm**
   Focus: Hearing, language, sound processing
6. **Perry the Parietal & The Body Map Bash**
   Focus: Touch, space, body awareness
7. **Hippocampus Hal & The Memory Maze**
   Focus: Memory, learning, recall
8. **Amy the Amygdala & The Overactive Alarm**
   Focus: Emotional alarm system
9. **Luna Lefty & Roger Righty's Mixed Up Masterpiece**
   Focus: Left/right hemispheres, teamwork
10. **Callie the Connector & The Brain Bridge**
    Focus: Corpus callosum, brain-to-brain communication

Series 2:
Cortex Castle (Frontal Lobe Executive Functions)

Series Overview: A lively castle of executive-function helpers. Kids practice planning, starting, stopping, focusing, flexible thinking, time skills, and self-talk, with Frontal Fred guiding thoughtful choices.

1. **Cortex Castle**
   Focus: Executive function overview
2. **Frontal Fred & The Decision Dilemma**
   Focus: Decision-making, prioritizing
3. **Impulsive Ian & The Button Box**
   Focus: Impulse control, pause
4. **Memory Molly & The Wandering Thoughts**
   Focus: Working memory, holding steps
5. **Focus Frannie & The Great Distraction Chase**
   Focus: Attention, refocusing
6. **Planella & The Checklist Parade**
   Focus: Planning, sequencing
7. **Flexi & The Changey-Changeroo Day**
   Focus: Cognitive flexibility
8. **Self-Talk Sally & The Thoughtful Echoes**
   Focus: Self-talk, emotional regulation
9. **Starter Stacy & Finisher Finn**
   Focus: Task initiation, task completion
10. M**ilo the Minute Keeper & Mae the Moment Seeker**
    Focus: Time awareness, pacing
11. **The Power of Pausing with Penny**
    Focus: Pause skills, regulation

Series 3:
Busy Brain Café (Neurodivergent Affirmations)

Series Overview: A welcoming café where different brain styles are celebrated. These stories affirm neurodivergent experiences while teaching self-understanding, communication, and compassion for others.

1. **Nathan Neuron & The Bouncy Brain**
   Focus: ADHD metaphor, movement, attention
2. **Sara Sprinkle the Sensa & The Super Sensitive Day**
   Focus: Sensory sensitivity, overload tools
3. **Overwhelmed Owen & The Senses Overload**
   Focus: Sensory overload, coping strategies
4. **Interoceptive Ivy & The Inside Signals**
   Focus: Interoception, body signals
5. **Zane's Zigzag Thoughts & The Path to Peace**
   Focus: Anxiety, grounding
6. **The Busy Brain Café**
   Focus: Neurodivergent inclusion, different processing styles
7. **The Listening Light**
   Focus: Nonverbal communication, understanding
8. **The Language of Love**
   Focus: Connection, communication, compassion
9. **Trying Theo & The How-To Hiccup**
   Focus: Persistence, frustration tolerance
10. **Drowsy Drew & The Nighttime Repair**
    Focus: Sleep, brain repair
11. **Burned-Out Bonnie & The Rebuild**
    Focus: Burnout, rest, recovery

Series 4:
Steady Systems Squad (Regulation & Growth)

Series Overview: A squad that explains how brains regulate, reset, rest, and grow stronger over time, including stress waves, sleep crews, sensory filtering, and rebuilding after hard days.

1. **The Steady Systems Squad**
   Focus: Regulation systems overview
2. **Timothy Thalamus & The Traffic Tower**
   Focus: Sensory filtering, attention routing
3. **Danny Dimmer & Wake-Up Wonda**
   Focus: Arousal and alertness
4. **The Start, Stop, and Switch Station**
   Focus: Initiation, inhibition, shifting
5. **The Night Shift & The Sleepy Crew**
   Focus: Sleep systems, restoration
6. **The Stress Wave & The Return to Calm**
   Focus: Stress response, recovery
7. **The Brain Growth Gathering**
   Focus: Brain growth and development
8. **Paul's Practice Path**
   Focus: Neuroplasticity
9. **Daisy's Daydream Network**
   Focus: Daydream network, imagination
10. **Remy's Rebuild Crew**
    Focus: Recovery and healing

NEXTGEN LEARNERS PRESENTS

Think Like a Brain!™ Collection

Perry the Parietal & The Body Map Bash

Knowing Where Your Body Is and How It Moves

BOOK 6

WRITTEN &
ILLUSTRATED BY:

MAISEL MCLAULA

NEXTGEN LEARNERS PRESS™
CINCINNATI, OHIO

This book is published by NextGen Learners Press, an imprint of NextGen Learners LLC.

Maisel McLaula is a pen name used by the author of this work.

First Edition: 2026
Printed in the United States of America
ISBN: 978-1-972860-05-2

This book is intended for educational and informational purposes only. It is not intended to replace professional medical, psychological, or therapeutic advice, diagnosis, or treatment.
Always seek the advice of a qualified professional with any questions regarding a medical or mental health condition.

Artwork and visual elements in this book were created using a combination of original artwork, with enhanced original direction, digital illustration tools, and AI-assisted creative technologies.

A Note for My Readers:

Hi friend,
Your brain is incredible. It thinks, feels, remembers, solves, dances, creates, and so much more!
This series is here to help you understand how your brain works, and to cheer you on as you grow your own thinking superpowers.

Every part of your brain has a job.
Sometimes they get mixed up. That's okay!

When you learn what each part does, you can work with your brain instead of feeling like it's working against you.
So read, giggle, learn, and maybe even draw your own Brainy Bunch when you're done.

I'm cheering you on with every rhyme.

-Maisel McLaula

Think Like a Brain!™

Brain Anatomy Collection

Explore the castle inside your head! Meet each room and its brilliant resident, from Frontal Fred to Benny the Brainstem. Whimsy, wonder, and neurons await!

For Leo & Maylee,
Whose strength, determination, and love of movement show how powerful growing bodies can be when limits are tested with joy and persistence.

And...

To the beautifully wired, the curiously wired, and the wonderfully neurodivergent minds, may you feel at home in your body.

You are exactly what a great brain looks like.

-Maisel McLaula

In the middle room where the senses all play,
Lived Perry the Parietal, mapping the way.
He knew where your fingers and elbows belonged,
And could tell if your sleeve felt itchy or wrong.

He tracked where your legs were, your nose and your toes,
Even when blindfolded or tangled in clothes.

But sometimes you grow, and your body feels new,
And the map that once fit needs updating too.

One Dance Bash Day morning, things went askew,
Perry misplaced the where, the what, and the who.
He blinked and he paused, then thought, “Oh, I see,
My body feels off, because I grew suddenly!”

He waved at a wall. High-fived a plant.
He tried to sit down, but missed by a slant.

The floor felt too tippy. The chair felt bizarre.
His elbow said "Here!" while his knee said "Where are we, so far?"

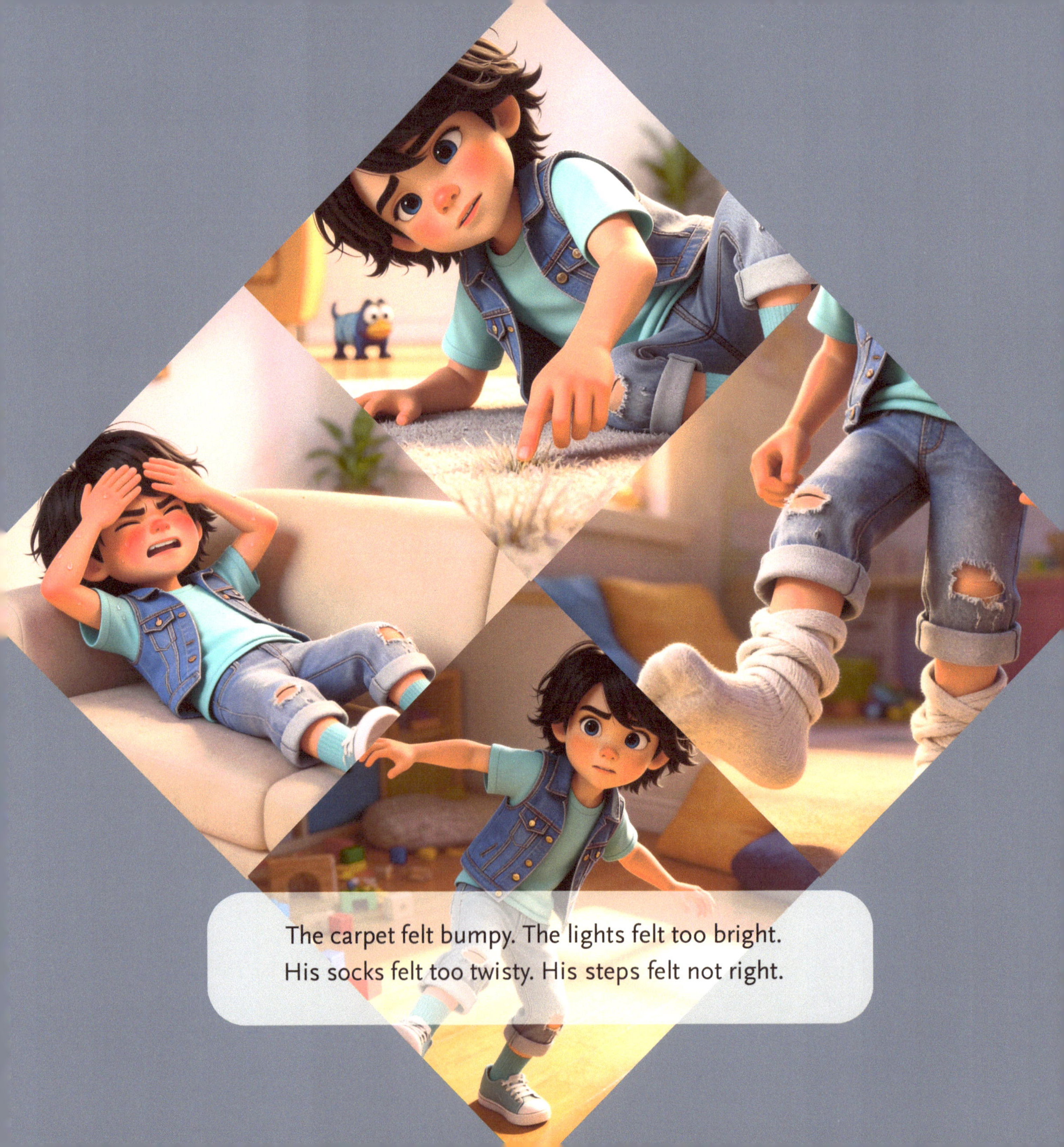

The carpet felt bumpy. The lights felt too bright.
His socks felt too twisty. His steps felt not right.

His left felt like right. His right felt like air.
Perry frowned softly. "My map needs repair."

The Body Map Bash was just hours away,
And Perry's own body felt lost in dismay.

He tripped over shoelaces, bumped into Fred,
Bonked into Bella, then bounced off a bed.

"Where is my knee? What happened to space?
I can't find my rhythm, I'm all over the place!"

Fred knelt down calmly and said with a smile,
"Let's slow your whole body and feel it a while.

We'll notice your parts from your head to your feet,
And help your brain learn where your body should meet."

They pressed rice and beans between fingers and toes.
They stretched, then they squished, then they wiggled in rows.

Warm socks hugged ankles. Cold bells chimed near.
Each signal said clearly, 'Your body is here.'

Perry felt knees. He noticed his heels.
He traced where his shoulders rolled, rested, and wheeled.

His map grew more certain with every small clue,
As the picture inside him began to feel true.

They practiced left, right with hopscotch and lines.
Eyes closed, Perry traced shapes and designs.

He tossed rings on bottles, then clapped with a pause,
Listening closely to signals from fingers and paws.

The Bash Day returned, and the music began,
With dancers and dashers and a popcorn stand.

Perry led games with a hop, spin, and flash,
Proudly presenting the Body Map Bash.

He showed where the hand meets the wrist with a charm,
How feet tell your brain when they're chilly or warm.

He pointed to elbows and even the spine,
Marching in rhythm, in order, in line.

The Brainy Bunch clapped and joined in with joy,
Each person, each grown-up, each girl and each boy.

They noticed their bodies in brand-new ways,
And Perry just grinned, "That's mapping, hooray!"

Now Perry still has a few bumpy growth days,
But he knows how to notice in sensory ways.

With texture and movement and visual check,
He tunes in his map from his feet to his neck.

So if you feel wonky or trip on your shoe,
Or you can't quite sense what your body should do,

Take time to sense, to stretch, and to tap,
And gently redraw your own body map.

Perry's Body Map Tools

- Use touch, stretch, and balance play
- Try hopscotch, yoga, or mirror relay
- Use textured paths or beanbags for aim
- Close eyes and notice where you feel the same

What Is the Parietal Lobe?

The parietal lobe helps your brain sense touch, space, and direction. It tells you where your body is and what you're feeling. It works with vision and movement to create a body map.

Try It Yourself!

- Draw a map of your hand and label each part
- Try a mirror walk or sensory chart
- Place stickers on toes and guess where they land
- Stretch your arms wide and make a Body Bash Band

Congratulations!

You've marched with Perry through sensation and space,
You've felt your own rhythm and found your own place.
Now your body map's clearer from finger to toe,
And your parietal lobe is ready to go.

Thank You to My Readers

Thank you for reading this book.
Whether you're a parent, a teacher,
a therapist, or a brilliant young
brain explorer,
your interest in understanding
how the mind works
means the world.

Sharing these books
means spreading compassion and
curiosity.

You are the reason
these stories come alive.
Thank you for believing in the
power of playful learning.

Stay curious. Stay kind.
Stay brainy!

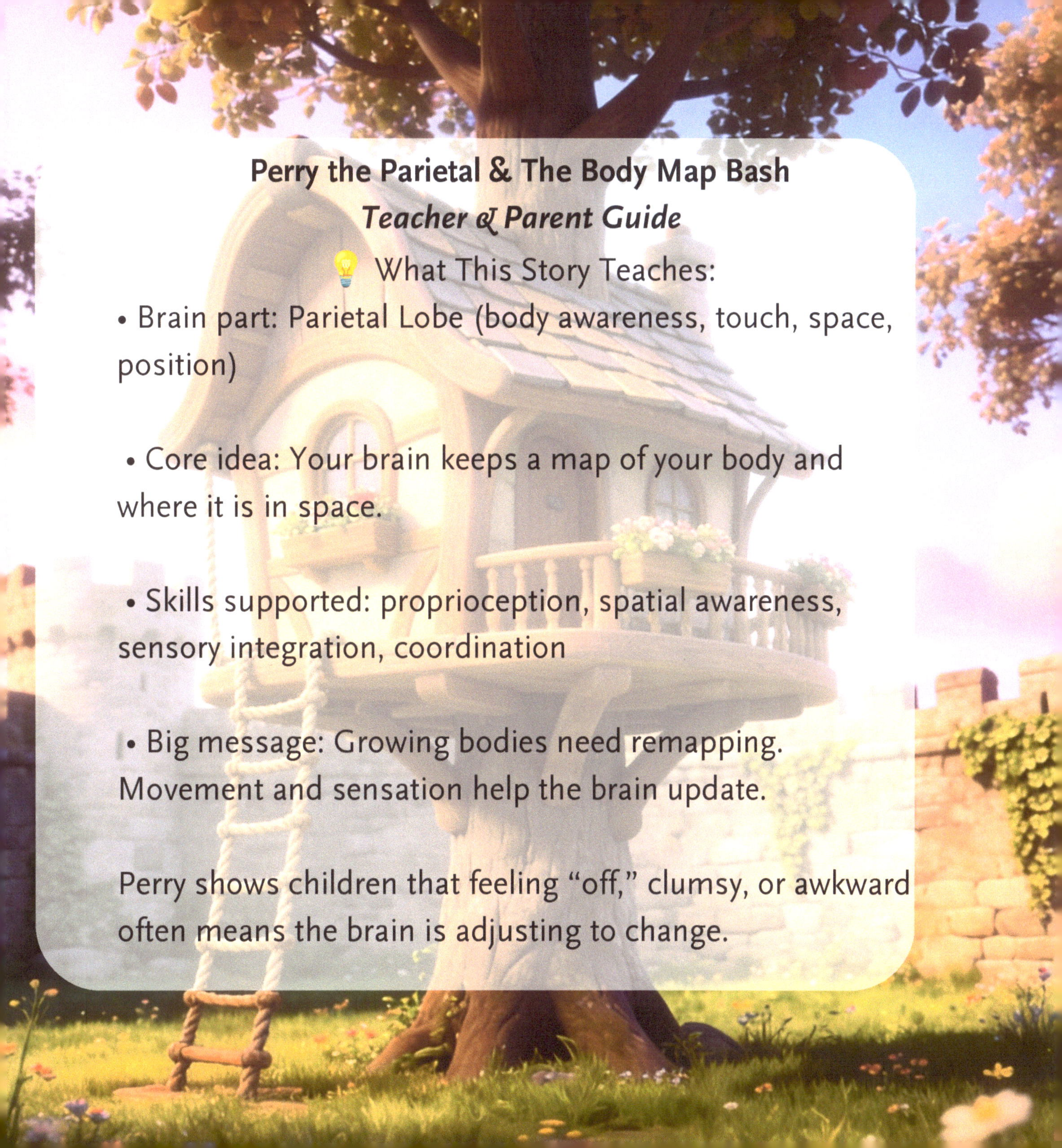

Perry the Parietal & The Body Map Bash

Teacher & Parent Guide

💡 What This Story Teaches:

- Brain part: Parietal Lobe (body awareness, touch, space, position)

- Core idea: Your brain keeps a map of your body and where it is in space.

- Skills supported: proprioception, spatial awareness, sensory integration, coordination

- Big message: Growing bodies need remapping. Movement and sensation help the brain update.

Perry shows children that feeling "off," clumsy, or awkward often means the brain is adjusting to change.

🧠 Key Concepts for Grown-Ups

Perry represents the parietal lobes, located near the top middle of the brain. These regions combine touch, movement, vision, and spatial information to create an internal body map. This map helps children know where their body parts are without looking.

When the parietal system is developing or overloaded, children may:

- Bump into things
- Trip or fall often
- Struggle with personal space
- Appear clumsy during growth spurts
- Seek deep pressure or avoid touch
- Have trouble with left/right
- Feel "not quite in their body"

Important foundations:

- Growth spurts temporarily disrupt body maps.
- Sensory input helps rebuild spatial awareness.
- Deep pressure and movement organize the nervous system.
- Body awareness improves with playful repetition.

This story helps children understand that:

- Feeling awkward is normal.
- Their brain is updating its map.
- Movement helps everything reconnect.

🗣 Discussion Starters

- What felt confusing for Perry?
- How did Perry help his body feel clearer again?
- When do you feel wobbly or unsure in your body?
- What movements help you feel grounded?
- Can you notice where your feet are right now?

Encourage children to describe sensations, not judge them.

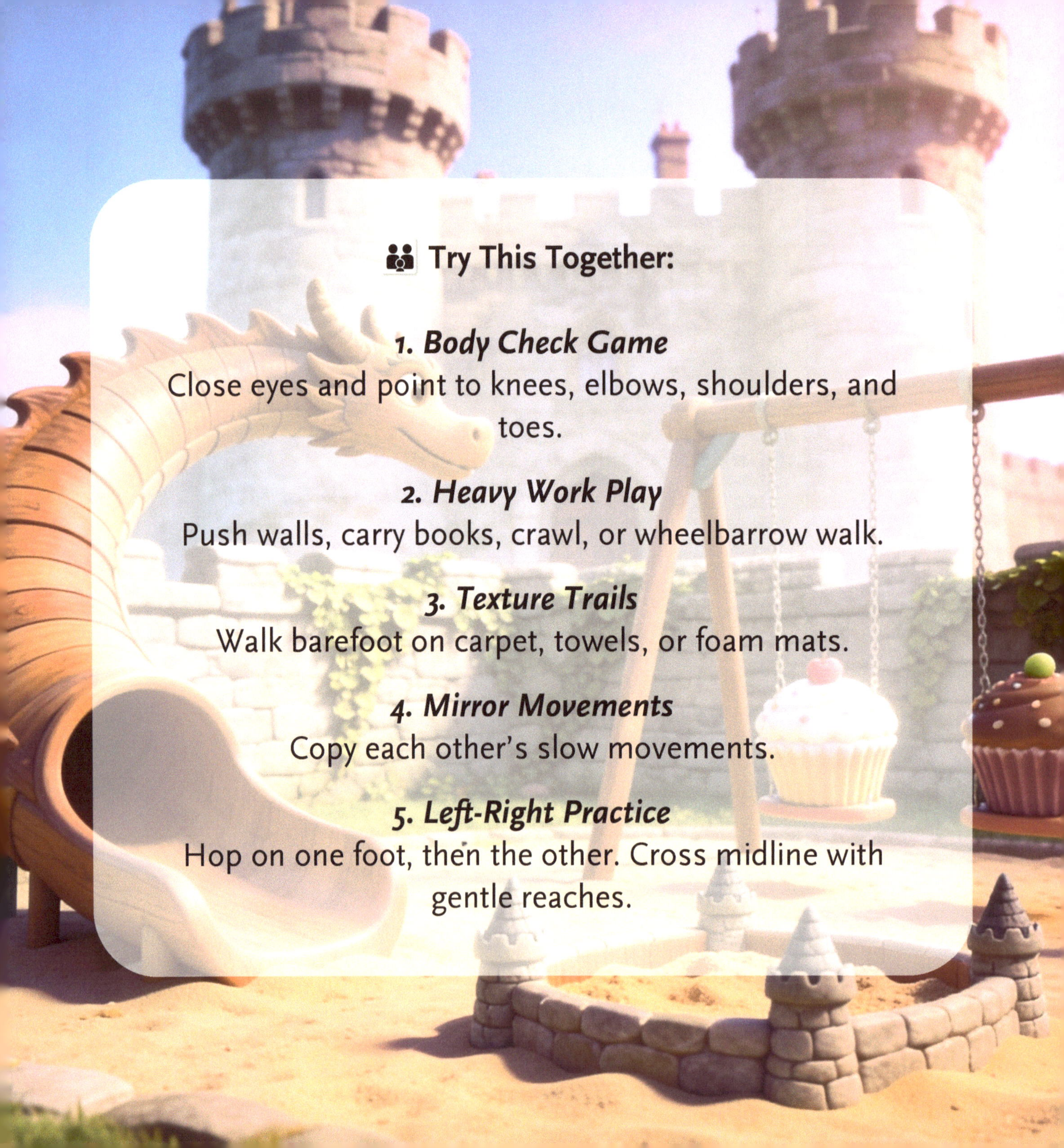

👨‍👩‍👧 Try This Together:

1. Body Check Game

Close eyes and point to knees, elbows, shoulders, and toes.

2. Heavy Work Play

Push walls, carry books, crawl, or wheelbarrow walk.

3. Texture Trails

Walk barefoot on carpet, towels, or foam mats.

4. Mirror Movements

Copy each other's slow movements.

5. Left-Right Practice

Hop on one foot, then the other. Cross midline with gentle reaches.

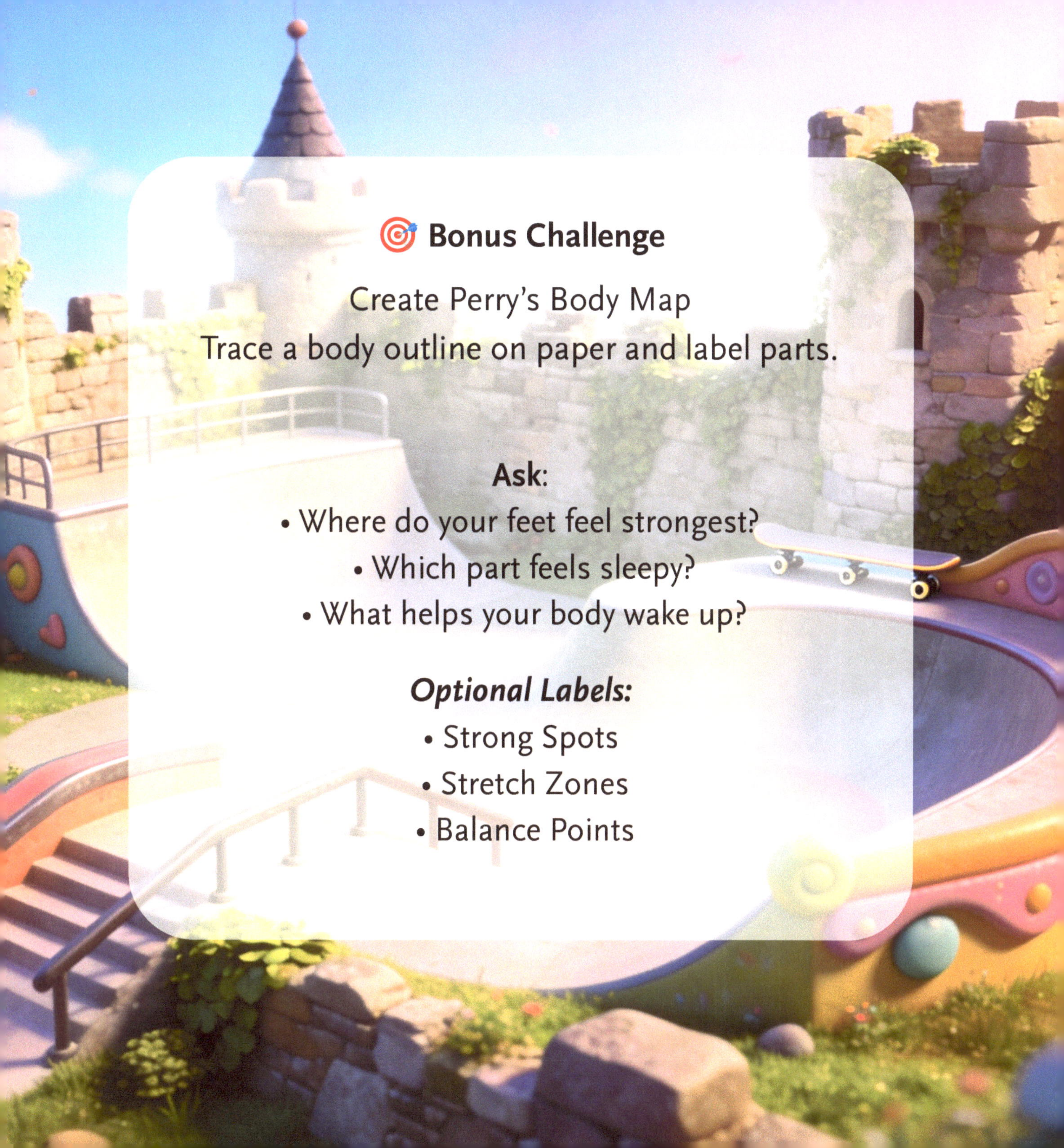

🎯 Bonus Challenge

Create Perry's Body Map

Trace a body outline on paper and label parts.

Ask:

- Where do your feet feel strongest?
- Which part feels sleepy?
- What helps your body wake up?

Optional Labels:

- Strong Spots
- Stretch Zones
- Balance Points

Playful Extension:

Build-a-Body Path

Create a simple movement path using floor spots, pillows, or tape:

- Step here
- Reach there
- Crawl under
- Balance across

Ask children to notice how they move their bodies from one spot to the next.

🛠️ Perry's Body Toolbox:

Use these tools when children struggle with coordination or spatial awareness:

- Offer frequent movement breaks
- Include pushing, pulling, and carrying
- Use weighted items when appropriate
- Provide clear boundaries for personal space
- Practice bilateral activities
- Normalize clumsiness during growth

Helpful phrases:

- "Your body map is updating."
- "Let's give your muscles some work."
- "Movement helps Perry reconnect."

Avoid labeling children as clumsy. Their brains are learning.

❤️ Final Note for Grown-Ups

Perry reminds us that bodies grow faster than brains can always keep up.

When children seem awkward or disconnected, their nervous systems are often simply recalibrating.

Movement, pressure, and playful awareness help the brain redraw its map.

Every stretch, crawl, and wiggle builds connection.
Every sensation teaches the brain where it lives.

The body map always finds its way back.

Resources & GROWTH

Resources:

(Parietal lobe + body awareness, spatial processing, touch)

- National Institute of Neurological Disorders and Stroke (NINDS)
- Cleveland Clinic
- MedlinePlus (U.S. National Library of Medicine)
- Encyclopedia Britannica
- NINDS (NIH). Brain Basics: Know Your Brain (parietal lobe/somatosensory function).
- Mayo Clinic. How your brain works (lobes and functions overview).
- NINDS (NIH). Brain Basics: Know Your Brain / Brain Basics resources.
- Cleveland Clinic. Brain: parts and function (overview pages).
- CASEL. CASEL's SEL Framework (self-awareness, self-management, social awareness, relationship skills, responsible decision-making).
- CDC. Child Development: Positive Parenting Tips and developmental guidance by age.

Growth:

(If you or someone you know wants to strengthen body awareness or spatial skills)

- Understood.org
  - Search: spatial awareness
- American Occupational Therapy Association (AOTA)
  - Look for: body awareness

Acknowledgments:

To my students: past, present, and future:
You taught me more than any textbook ever could. By watching how you explored, questioned, struggled, and discovered, I learned how minds truly grow. Every "aha" moment in these pages began with you.

To the parents, teachers, and guides walking beside children every day:
Thank you for listening, for advocating when things feel complicated, and for believing that understanding a child's inner world truly matters. Your patience and care shape futures in ways you may never fully see.

To my husband, my best friend, and my constant partner in learning:
Thank you for walking beside me, believing in this work, and holding space for creativity to exist. Sharing curiosity, growth, and creation with you has been one of the greatest honors of my life.

To the great thinkers who shaped how I see the world: including Dr. Seuss, Albert Einstein, Leonardo da Vinci, Stephen Hawking, Ludwig van Beethoven, just to name a few, and my English teacher, Mrs. Godfrey:
Your curiosity, courage, and imagination reminded me that learning is limitless, creativity is powerful, and questions matter. Mrs. Godfrey, your encouragement, belief in me, and love of words sparked my writing journey. If not for you, I may never have picked up a pen.

And to my publisher, NextGen Learners™:
It was by fate that we met, and by inspiration that we now move forward. Together, we build something meaningful. Together, we leave this legacy.

Lastly, to my own brain:
Complex, resilient, ever-adapting, and endlessly learning. Thank you for carrying stories, holding wonder, and reminding me that growth is always possible.

Maisel McLaula is the author of the Think Like a Brain!™ children's book collection, a story-first series that blends neuroscience, imagination, and emotional safety for young readers.

Drawing from decades of experience supporting children and families, her work centers on a simple but powerful belief: children deserve to understand and accept who they are, including how their brains work and why they make the choices they do.

Through playful characters and gentle metaphors, Maisel's stories help children discover the "why" behind their thoughts, feelings, and actions, turning confusion into clarity and self-judgment into self-compassion. By understanding their inner systems, children gain the power to respond with kindness toward themselves and to recognize those same patterns in others.

Maisel believes that when children feel safe inside their own minds, empathy grows naturally. The Think Like a Brain!™ collection offers a foundation for confidence, understanding, and care that grows with the child.

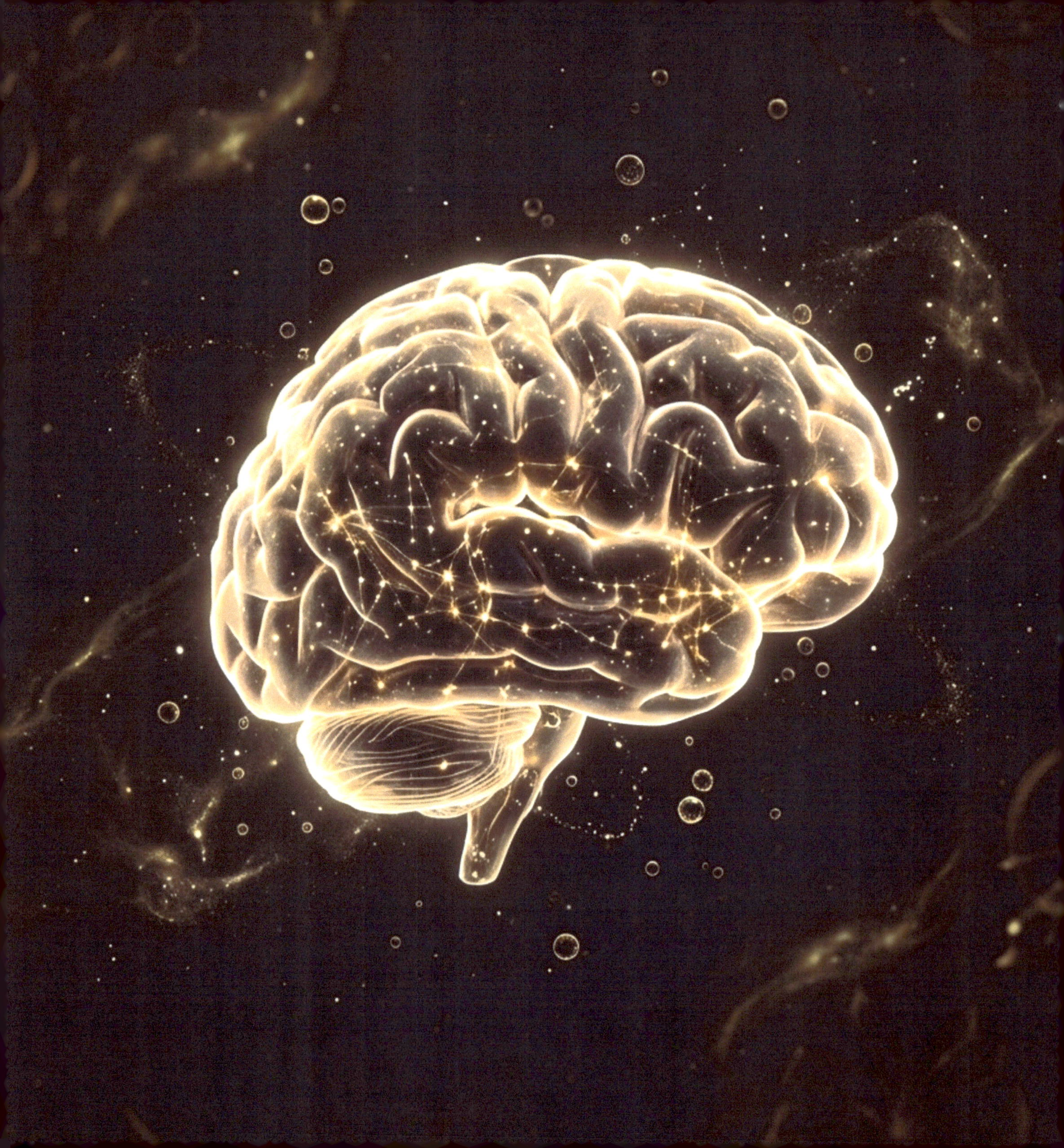

ABOUT THE PUBLISHER

NextGen Learners LLC is an educational publishing company dedicated to helping children understand how they think, feel, learn, and grow.

Through imaginative storytelling, neuroscience-informed concepts, and real-world application, NextGen Learners™ creates books and resources that support emotional development, critical thinking, self-awareness, and confidence in young minds.

Our work is designed not only for children, but for the parents, educators, caregivers, and professionals who walk beside them. Each book serves as a bridge, helping adults better understand the child in front of them, while helping children better understand themselves.

By creating shared language and meaningful connection points, our stories foster deeper relationships, reduce misunderstanding, and encourage acceptance, both within oneself and toward others.

We believe every child's mind is uniquely wired, and that understanding those differences is the foundation for compassion, growth, and lifelong learning.
With multiple collections designed to support cognitive, emotional, and developmental growth, NextGen Learners provides tools that children can carry with them, and tools that adults can use to guide them with clarity, patience, and purpose.

NG NEXTGEN LEARNERS PRESS

MORE TO EXPLORE

The Brainy Bunch is just the beginning.
More stories, tools, and brain-friendly resources are on the way! Designed to help children, parents, and educators better understand how the brain works in everyday life.

FREE PRINTABLES

We've created printable pages to go along with this book, including:

-Reading Guides
-Brain helper visuals
-Character Cards

These are available for free as part of our growing resource library.

Our full website experience is built to include:

-Downloadable activity packs
-Parent & teacher guides
-Expanded learning tools
-Additional collections and topics

STAY CONNECTED

We love hearing how children connect with these stories!

If there is a topic, feeling, or challenge you would like to see explored in future books, we would love to hear from you!

Contact us: Inspire@NextGenLearners.com
NextGenLearners.com

Your ideas help shape future stories and support the growth of this collection.

This collection grows with you. New resources and stories are added regularly.

Maisel McLaula loves hearing from the tiny humans who read her books.

Did a story make you giggle, spark a big idea, or inspire you to draw your very own character? She would be delighted to see it! Send your letters and artwork to:

NextGen Learners
Attn: Maisel McLaula
8666 Beechmont Ave #1063
Cincinnati, OH 45255
Every letter is read with a smile, and your artwork just might make her whole day.

OPEN BRAIN

www.ingramcontent.com/pod-product-compliance
Lightning Source LLC
Chambersburg PA
CBHW041321210926
51847CB00002B/18
9781972860052